VECTORS
IBH Topic 4

Steven Clarke
United Nations International School
New York

This second edition covers all of the material required for Topic 4 (Vectors) in the new syllabus of IB Higher Level Mathematics.

The book can be used for self-study, as numerous worked examples are provided .

If you find any typographical errors, or incorrect answers, please let me know at   sclarke@unis.org

SECOND EDITION

ISBN   978-1478393566

Contents

4

Presumed Knowledge :

Basic 3-D coordinate geometry

Trigonometry

Bearings

Notation:

A(x, y, z)	the point A in the Cartesian plane with coordinates x, y and z
[AB]	the line segment with end points A and B
$\hat{A}$	the angle at A
C$\hat{A}$B	the angle between [CA] and [AB]
Δ ABC	the triangle with vertices A, B and C
a	the position vector $\overrightarrow{OA}$
i, j, k	unit vectors in the directions of the Cartesian coordinate axes
$\lvert a \rvert$	the magnitude of a
$\lvert \overrightarrow{AB} \rvert$	the magnitude of $\overrightarrow{AB}$
$a \cdot b$	the scalar product of a and b
$a \times b$	the vector product of a and b

Introduction

A vector has magnitude and direction. A scalar has only magnitude.
'55 km/hr' is a scalar. '55 km/hr North' is a vector.

The scalar '55 km/hr' is called a 'speed'.
The vector '55 km/hr North' is called a 'velocity'.

$\overrightarrow{AB}$ represents a vector that begins at point A and ends at point B.

$-\overrightarrow{AB} = \overrightarrow{BA}$

$\overrightarrow{OA}$ represents a vector that begins at the origin O and ends at point A.
Vectors which begin at the origin O are called **position vectors**.

Vectors can also be represented by single letters, for example, **a** or **a**

If A has coordinates $(3, 4)$ then the vector $\overrightarrow{OA}$ can be written as $\begin{pmatrix} 3 \\ 4 \end{pmatrix}$

This indicates a line segment starting at the origin $(0, 0)$ and ending at
the point $(3, 4)$

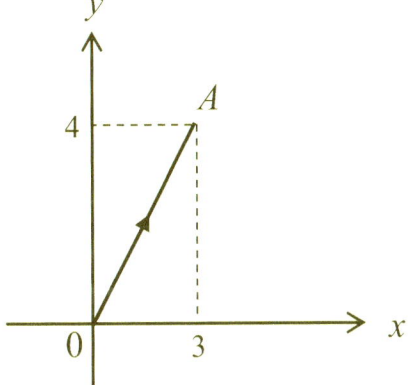

The magnitude of vector $\overrightarrow{OA}$ is the length of the line segment (in this case: 5 units). It is represented by $\left|\overrightarrow{OA}\right|$

The direction of vector $\overrightarrow{OA}$ is $53.1°$ from the positive x-axis.

In general, the magnitude of $\begin{pmatrix} x \\ y \end{pmatrix}$ is $\sqrt{x^2 + y^2}$

and the direction is found by using simple trigonometry.

In 3-dimensions, the magnitude of $\begin{pmatrix} x \\ y \\ z \end{pmatrix}$ is $\sqrt{x^2 + y^2 + z^2}$

and the direction is defined by using "direction cosines" (this is not required for IBH).

Example

If $P = (2, 3, 6)$ then $\overrightarrow{OP} = \begin{pmatrix} 2 \\ 3 \\ 6 \end{pmatrix}$

and the magnitude (length) of $\overrightarrow{OP}$ is given by

$$\left|\overrightarrow{OP}\right| = \sqrt{2^2 + 3^2 + 6^2}$$
$$= 7$$

Parallel vectors have the same (or reverse) direction.

a, $3a$, $8a$, $-a$, $-4a$, ka $(k \in R)$ are all parallel vectors.

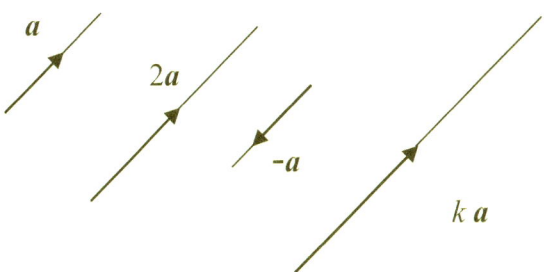

In general , if a and b are parallel vectors then $a = kb$
for some value of $k \in R$

Example

$\begin{pmatrix} 4 \\ 2 \end{pmatrix}$ and $\begin{pmatrix} 6 \\ x \end{pmatrix}$ are parallel vectors. Find x.

Parallel $\Rightarrow$ $\begin{pmatrix} 4 \\ 2 \end{pmatrix} = k \begin{pmatrix} 6 \\ x \end{pmatrix}$

$\Rightarrow$ $4 = 6k$ and $2 = kx$

$\Rightarrow$ $k = \dfrac{2}{3}$ and hence $x = 3$

Addition of vectors

$$\overrightarrow{AB} + \overrightarrow{BC} = \overrightarrow{AC}$$

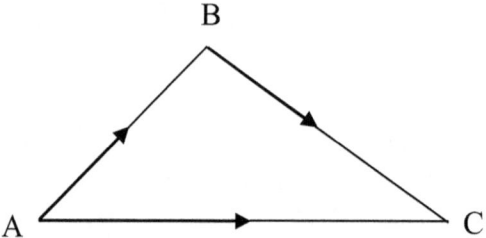

In other words, when moving from A to B to C, the *displacement* is the same as when moving from A to C.

Conversely $\overrightarrow{AB}$ = $\overrightarrow{AO} + \overrightarrow{OB}$

= $-\overrightarrow{OA} + \overrightarrow{OB}$

= $\overrightarrow{OB} - \overrightarrow{OA}$

Example

Given $A = (3, 4)$ and $B = (8, 5)$ find $\overrightarrow{AB}$

$$\overrightarrow{AB} = \overrightarrow{OB} - \overrightarrow{OA}$$

$$= \begin{pmatrix} 8 \\ 5 \end{pmatrix} - \begin{pmatrix} 3 \\ 4 \end{pmatrix}$$

$$= \begin{pmatrix} 5 \\ 1 \end{pmatrix}$$

Example

Given $A = (1, 2, 3)$ and $B = (2, 3, 6)$ find $\left|\overrightarrow{AB}\right|$

$$\overrightarrow{AB} = \overrightarrow{OB} - \overrightarrow{OA}$$

$$= \begin{pmatrix} 2 \\ 3 \\ 6 \end{pmatrix} - \begin{pmatrix} 1 \\ 2 \\ 3 \end{pmatrix}$$

$$= \begin{pmatrix} 1 \\ 1 \\ 3 \end{pmatrix}$$

Hence $\left|\overrightarrow{AB}\right| = \sqrt{1^2 + 1^2 + 3^2}$

$$= \sqrt{11}$$

Simplifying vectors:

$$\overrightarrow{AB} + \overrightarrow{BC} + \overrightarrow{CD} = \overrightarrow{AD}$$

$$\overrightarrow{AB} + \overrightarrow{CA} = \overrightarrow{CA} + \overrightarrow{AB}$$
$$= \overrightarrow{CB}$$

$$\overrightarrow{PQ} - \overrightarrow{RQ} = \overrightarrow{PQ} + \overrightarrow{QR}$$
$$= \overrightarrow{PR}$$

The following diagram illustrates $a + b$

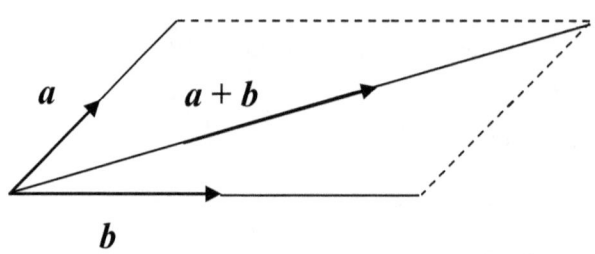

The following diagram illustrates $a - b$

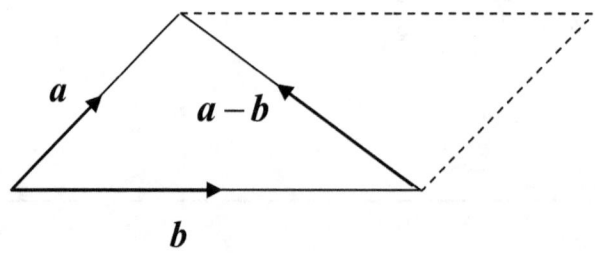

If $|a - b| = |a + b|$ then the parallelogram becomes a rectangle and hence vectors a and b will be perpendicular.

If $|a| = |b|$ then the parallelogram becomes a rhombus and hence vectors $a - b$ and $a + b$ will be perpendicular.

Unit Vectors

By definition, unit vectors have a length of ONE.

Unit vectors along the x, y, z axes are written as $\boldsymbol{i}, \boldsymbol{j}, \boldsymbol{k}$ respectively.

Hence $\qquad \boldsymbol{i} = \begin{pmatrix} 1 \\ 0 \\ 0 \end{pmatrix} \qquad\qquad \boldsymbol{j} = \begin{pmatrix} 0 \\ 1 \\ 0 \end{pmatrix} \qquad\qquad \boldsymbol{k} = \begin{pmatrix} 0 \\ 0 \\ 1 \end{pmatrix}$

Vectors can be written using $\boldsymbol{i}, \boldsymbol{j}, \boldsymbol{k}$

For example $\qquad \begin{pmatrix} 3 \\ 4 \\ 5 \end{pmatrix} = 3\boldsymbol{i} + 4\boldsymbol{j} + 5\boldsymbol{k}$

$\begin{pmatrix} 3 \\ 4 \\ 5 \end{pmatrix}$ is called a column vector.

This notation is much easier to use than $3\boldsymbol{i} + 4\boldsymbol{j} + 5\boldsymbol{k}$

Scalar Product

The scalar product (or dot product) of two vectors $\boldsymbol{a}$ and $\boldsymbol{b}$ is given by

$$\boldsymbol{a} \cdot \boldsymbol{b} = |a|\ |b|\ \cos\theta$$

where θ is the angle between the two vectors.

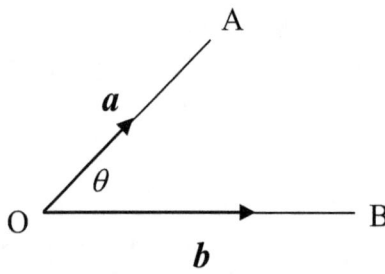

Note the direction of the vectors: both away from the point O.

For unit vectors $\boldsymbol{i}, \boldsymbol{j}, \boldsymbol{k}$ we have

$$\boldsymbol{i} \cdot \boldsymbol{i} = |i|\ |i|\ \cos 0°$$

$$= 1$$

$$\boldsymbol{i} \cdot \boldsymbol{j} = |i|\ |j|\ \cos 90°$$

$$= 0$$

Similarly $\quad \boldsymbol{i} \cdot \boldsymbol{i} = \boldsymbol{j} \cdot \boldsymbol{j} = \boldsymbol{k} \cdot \boldsymbol{k} = 1$

$$\boldsymbol{i} \cdot \boldsymbol{j} = \boldsymbol{i} \cdot \boldsymbol{k} = \boldsymbol{j} \cdot \boldsymbol{k} = 0$$

Example

Given $\overrightarrow{OA} = 2i + 3j$ and $\overrightarrow{OB} = 4i + 5j$, calculate $\overrightarrow{OA} \cdot \overrightarrow{OB}$

$$\overrightarrow{OA} \cdot \overrightarrow{OB} = (2i + 3j) \cdot (4i + 5j)$$

$$= 8\,i \cdot i + 10\,i \cdot j + 12\,j \cdot i + 15\,j \cdot j$$

$$= 8 + 0 + 0 + 15$$

$$= 23$$

A short-cut method is shown below:

$$\overrightarrow{OA} \cdot \overrightarrow{OB} = \begin{pmatrix} 2 \\ 3 \end{pmatrix} \cdot \begin{pmatrix} 4 \\ 5 \end{pmatrix}$$

$$= 8 + 15$$

$$= 23$$

Example

Calculate the scalar product $\begin{pmatrix} 2 \\ 3 \\ 6 \end{pmatrix} \cdot \begin{pmatrix} 4 \\ 1 \\ 3 \end{pmatrix}$

$$\begin{pmatrix} 2 \\ 3 \\ 6 \end{pmatrix} \cdot \begin{pmatrix} 4 \\ 1 \\ 3 \end{pmatrix} = 8 + 3 + 18$$

$$= 29$$

Example

Calculate the angle between the vectors $\begin{pmatrix} 2 \\ 3 \\ 6 \end{pmatrix}$ and $\begin{pmatrix} 4 \\ 1 \\ 3 \end{pmatrix}$

Using $\quad \boldsymbol{a} \cdot \boldsymbol{b} = |\boldsymbol{a}|\,|\boldsymbol{b}|\,\cos\theta \quad$ we have

$$8 + 3 + 18 = \sqrt{2^2 + 3^2 + 6^2}\;\sqrt{4^2 + 1^2 + 3^2}\;\cos\theta$$

$$29 = \sqrt{49}\,\sqrt{26}\,\cos\theta$$

Hence $\quad \cos\theta = \dfrac{29}{7\sqrt{26}}$

$$= 35.7^\circ$$

Properties of Scalar Product

1) Scalar product is commutative :

$$a \cdot b = |a| \; |b| \; \cos\theta$$
$$= |b| \; |a| \; \cos\theta$$

$$= b \cdot a$$

2) Scalar product is distributive :

$$a \cdot (b+c) = a \cdot b + a \cdot c$$

3) $(ka) \cdot b = k(a \cdot b)$ where k is a constant

4) $$a \cdot a = |a| \; |a| \; \cos 0$$

$$= |a|^{2}$$

5) If a and b are parallel vectors then the magnitude of $a \cdot b$ is given by $|a| \; |b|$

Exercise 1

1) Given $A = (3, 8, 2)$ $B = (5, 6, 9)$

    (a) find $\overrightarrow{AB}$

    (b) calculate $\left|\overrightarrow{AB}\right|$

2) Calculate the angle between the following vectors:

    (a) $\begin{pmatrix} 3 \\ 4 \\ 5 \end{pmatrix}$ and $\begin{pmatrix} 1 \\ 1 \\ 3 \end{pmatrix}$

    (b) $4i - 2j + 3k$ and $3i + j + k$

    (c) $\begin{pmatrix} -2 \\ 5 \end{pmatrix}$ and $\begin{pmatrix} 8 \\ -2 \end{pmatrix}$

3) Show that $\begin{pmatrix} -3 \\ 6 \\ 1 \end{pmatrix}$ and $\begin{pmatrix} 2 \\ 2 \\ -6 \end{pmatrix}$ are perpendicular.

4) Given $A = (2, 4, 7)$ $B = (3, 6, 9)$ $C = (1, 6, 1)$

   (a) calculate (i) $\left|\overrightarrow{AB}\right|$ (ii) $\left|\overrightarrow{AC}\right|$

   (b) calculate angle BAC correct to 2 d.p.

   (c) hence calculate the area of triangle ABC.

5) Show that $\begin{pmatrix} \cos\theta \\ \sin\theta \end{pmatrix}$ is a unit vector

6) Find a unit vector in the same direction as

   (a) $3i + 4j$

   (b) $\begin{pmatrix} 2 \\ 3 \\ 6 \end{pmatrix}$

   (c) $3i + j + k$

7) $\overrightarrow{PQ}$ is parallel to $2i - 6j + 3k$ and in the same direction.

If $\overrightarrow{PQ}$ has length 14 then write $\overrightarrow{PQ}$ in the form $ai + bj + ck$

where $a, b, c \in R$

8) Given $a = \begin{pmatrix} 4 \\ k \end{pmatrix}$ and $b = \begin{pmatrix} 3 \\ 2 \end{pmatrix}$

(a) calculate the value of k if a and b are perpendicular

(b) calculate the value of k if a and b are parallel.

9) Given $a = \begin{pmatrix} 3 \\ 5 \\ k \end{pmatrix}$ and $b = \begin{pmatrix} 5 \\ 7 \\ 2 \end{pmatrix}$

calculate the value of k if a and b are perpendicular.

10) If $|a + b| = |a - b|$ what must be true about vectors a and b ?

Vector Line Equations

In Cartesian form, the equation of a straight line can be written as $y = mx + c$ where m is the slope of the line and $(0, c)$ is a point on the line.

In vector form, the equation of a straight line can be written as $\boldsymbol{r} = \boldsymbol{a} + \lambda \boldsymbol{b}$ where $\boldsymbol{b}$ is a vector representing the direction of the line, $\boldsymbol{a}$ is the position vector of a point on the line, and $\lambda \in \mathbb{R}$

In both systems the line is defined by knowing its direction (slope) and its location in space (by knowing a point on the line).

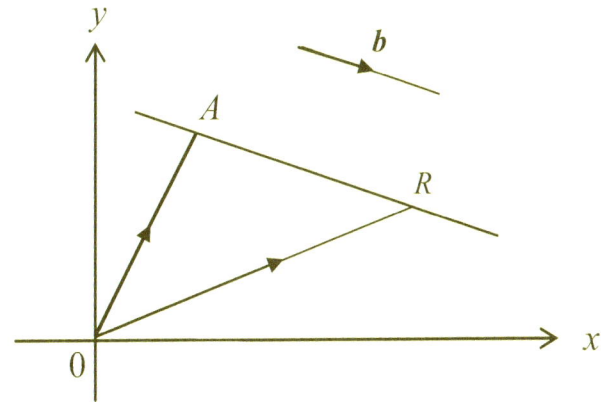

$$\overrightarrow{OR} = \overrightarrow{OA} + \overrightarrow{AR}$$

$$\overrightarrow{OR} = \overrightarrow{OA} + \lambda \boldsymbol{b}$$

$$\boldsymbol{r} = \boldsymbol{a} + \lambda \boldsymbol{b}$$

Note: Point A is a fixed point on the line.
       Point R is a general point on the line; it can move anywhere along the line.

Consider $\quad y = 2x + 6$

This is a straight line. It has slope 2 and passes through the point $(0, 6)$.

Consider $\quad r = \begin{pmatrix} 0 \\ 6 \end{pmatrix} + \lambda \begin{pmatrix} 1 \\ 2 \end{pmatrix}$

This is also a straight line. It has direction $\begin{pmatrix} 1 \\ 2 \end{pmatrix}$ and passes through the point

with position vector $\begin{pmatrix} 0 \\ 6 \end{pmatrix}$.

Since direction vector $\begin{pmatrix} 1 \\ 2 \end{pmatrix}$ has slope 2, and $\begin{pmatrix} 0 \\ 6 \end{pmatrix}$ is the position vector

of the point $(0, 6)$ it follows that $\quad y = 2x + 6 \quad$ and $\quad r = \begin{pmatrix} 0 \\ 6 \end{pmatrix} + \lambda \begin{pmatrix} 1 \\ 2 \end{pmatrix}$

are the same straight line.

This can easily be shown algebraically as follows:

r represents the vector $\begin{pmatrix} x \\ y \end{pmatrix}$ so we have $\begin{pmatrix} x \\ y \end{pmatrix} = \begin{pmatrix} 0 \\ 6 \end{pmatrix} + \lambda \begin{pmatrix} 1 \\ 2 \end{pmatrix}$

This gives $\quad x = 0 + \lambda \quad$ and $\quad y = 6 + 2\lambda$

Substituting gives $\quad y = 2x + 6$

The vector equation $r = \begin{pmatrix} 0 \\ 6 \end{pmatrix} + \lambda \begin{pmatrix} 1 \\ 2 \end{pmatrix}$ could also have been written as

$r = \begin{pmatrix} 1 \\ 8 \end{pmatrix} + k \begin{pmatrix} 3 \\ 6 \end{pmatrix}$ since $(1, 8)$ also lies on the line and $\begin{pmatrix} 3 \\ 6 \end{pmatrix}$ and $\begin{pmatrix} 1 \\ 2 \end{pmatrix}$

have the same direction (they are parallel vectors) .

In the form $y = mx+c$ the coordinate (x, y) is the general point on the line.

In the form $r = a + \lambda b$, r is the position vector $\begin{pmatrix} x \\ y \end{pmatrix}$ of the general point

(x, y) on the line.

Note: by giving λ numerical values we can find the set of all points that lie on the line.

In 3-dimensions $r = \begin{pmatrix} 2 \\ 3 \\ 4 \end{pmatrix} + \lambda \begin{pmatrix} 1 \\ 2 \\ 5 \end{pmatrix}$ represents a straight line passing

through $(2, 3, 4)$ and parallel to vector $\begin{pmatrix} 1 \\ 2 \\ 5 \end{pmatrix}$

In this case $r = \begin{pmatrix} x \\ y \\ z \end{pmatrix}$

Intersecting Lines

If two lines meet at the point (a, b, c)

then the position vector $\quad r = \begin{pmatrix} a \\ b \\ c \end{pmatrix} \quad$ must satisfy both line equations.

Example

Find the point of intersection of

$$r_1 = \begin{pmatrix} 1 \\ 2 \\ 1 \end{pmatrix} + \lambda \begin{pmatrix} 3 \\ 1 \\ 4 \end{pmatrix} \quad \text{and} \quad r_2 = \begin{pmatrix} -5 \\ 10 \\ 6 \end{pmatrix} + \mu \begin{pmatrix} 4 \\ -2 \\ 1 \end{pmatrix}$$

Step 1: Let $r_1 = r_2$

$$\begin{pmatrix} 1 \\ 2 \\ 1 \end{pmatrix} + \lambda \begin{pmatrix} 3 \\ 1 \\ 4 \end{pmatrix} = \begin{pmatrix} -5 \\ 10 \\ 6 \end{pmatrix} + \mu \begin{pmatrix} 4 \\ -2 \\ 1 \end{pmatrix}$$

Step 2: Rewrite as

$$1 + 3\lambda = -5 + 4\mu$$

$$2 + \lambda = 10 - 2\mu$$

$$1 + 4\lambda = 6 + \mu$$

Note that we have 3 equations, but only 2 unknowns

Step 3: Choose any 2 of the 3 equations and solve together simultaneously:

$$1 + 3\lambda = -5 + 4\mu$$

$$2 + \lambda = 10 - 2\mu$$

This gives $\lambda = 2,$ $\mu = 3$

Step 4: Check that these two values also satisfy the third equation.

(Since the question implies that there actually is a point of intersection, this step could be omitted. However it is always a good idea to check, just in case you have made a calculation error)

Step 5: Substitute either value for λ or μ back into the original equation to find the coordinates

$$r_1 = \begin{pmatrix} 1 \\ 2 \\ 1 \end{pmatrix} + 2 \begin{pmatrix} 3 \\ 1 \\ 4 \end{pmatrix}$$

$$r_1 = \begin{pmatrix} 7 \\ 4 \\ 9 \end{pmatrix}$$

Hence the point of intersection is $(7, 4, 9)$

Parallel and Skew lines

If there is no point of intersection then the two lines are either parallel or skew .

Parallel lines will have the same direction vector.

In the case of skew lines, **Step 4** in the example just shown will not work. (In other words, the values of λ and μ found from two of the equations will not satisfy the third equation.)

It is also possible to have infinite solutions if the two equations represent the same line. In this case the answer will be any point on the line.

Angle between two lines

Use scalar product to calculate the angle between the direction vectors of the two lines.

If the two lines are $\quad r = a + \lambda\, b \quad$ and $\quad r = p + \mu\, q$

then use $\quad b \cdot q \; = \; |b|\; |q|\; \cos\theta$

where θ is the required angle.

Example

Calculate the angle between the lines

$$r_1 = \begin{pmatrix} 1 \\ 2 \\ 1 \end{pmatrix} + \lambda \begin{pmatrix} 3 \\ 1 \\ 4 \end{pmatrix} \quad \text{and} \quad r_2 = \begin{pmatrix} -5 \\ 10 \\ 6 \end{pmatrix} + \mu \begin{pmatrix} 4 \\ -2 \\ 1 \end{pmatrix}$$

The direction vectors are $\begin{pmatrix} 3 \\ 1 \\ 4 \end{pmatrix}$ and $\begin{pmatrix} 4 \\ -2 \\ 1 \end{pmatrix}$

Hence $\begin{pmatrix} 3 \\ 1 \\ 4 \end{pmatrix} \bullet \begin{pmatrix} 4 \\ -2 \\ 1 \end{pmatrix} = \sqrt{3^2 + 1^2 + 4^2} \; \sqrt{4^2 + (-2)^2 + 1^2} \; \cos\theta$

$$12 - 2 + 4 = \sqrt{26} \; \sqrt{21} \; \cos\theta$$

$$\cos\theta = \frac{14}{\sqrt{26} \; \sqrt{21}}$$

$$\theta = 53.2°$$

Applications to kinematics

The vector line equation $r = a + t\,b$ can be used to represent motion.
The vector r gives the location of an object at time t.

Consider a ship sailing in a straight line such that its location is given by

$$r = \begin{pmatrix} 1 \\ 2 \end{pmatrix} + t \begin{pmatrix} 3 \\ 4 \end{pmatrix}$$

where $r = \begin{pmatrix} x \\ y \end{pmatrix}$ and t = time in hours.

It can be seen that when $t = 0$, $r = \begin{pmatrix} 1 \\ 2 \end{pmatrix}$

Hence the initial location of the ship (its starting point) is given
by the coordinates (1, 2).
This could represent 1 kilometre East and 2 kilometres North
from some origin O.

After one hour the ships location is (4, 6) (obtained by substituting $t = 1$)

Hence, in one hour, the ship has moved the distance between coordinates
(1, 2) and (4, 6). This is easily calculated as 5 kilometres.

Hence the speed of the ship is 5 km/hour

After two hours the ships location is (7, 10) (substitute $t = 2$)

A faster method for finding the speed of the ship is to calculate $|b|$

$$\text{Here } \; b = \begin{pmatrix} 3 \\ 4 \end{pmatrix} \quad \text{so} \quad |b| = \sqrt{3^2 + 4^2}$$

$$= 5 \;\; \text{km/hr}$$

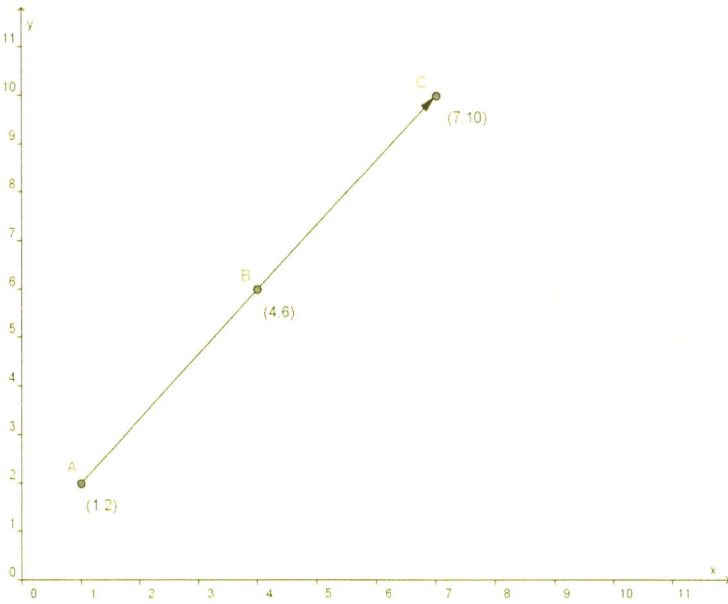

When $t = 0$ the ship is at the starting location A

The direction of travel of the ship is given by the direction of the vector $\overrightarrow{AB}$.

Since $\overrightarrow{AB} = \begin{pmatrix} 3 \\ 4 \end{pmatrix}$ the angle of direction will be $\tan^{-1}(\frac{4}{3})$ or $53.1°$

to the positive x-axis. This is written better as $N\,37°\,E$ or bearing $037°$

Example

A ship steams along path $\quad r = \begin{pmatrix} 3 \\ 3 \end{pmatrix} + t \begin{pmatrix} 10 \\ 24 \end{pmatrix}$

A yacht sails along path $\quad r = \begin{pmatrix} 31 \\ 43 \end{pmatrix} + t \begin{pmatrix} -4 \\ 4 \end{pmatrix}$

$\begin{pmatrix} 1 \\ 0 \end{pmatrix}$ represents one kilometre East of port P.

$\begin{pmatrix} 0 \\ 1 \end{pmatrix}$ represents one kilometre North of port P.

Time t is measured in hours after 12 noon.

1) *Find the speed of the ship.*

The direction vector for the motion of the ship is $\begin{pmatrix} 10 \\ 24 \end{pmatrix}$

Hence the speed $= \sqrt{10^2 + 24^2}$

$= 26 \ \text{km/hr}$

2) *Find the speed and bearing of the yacht.*

The direction vector for the motion of the yacht is $\begin{pmatrix} -4 \\ 4 \end{pmatrix}$

Hence the speed $= \sqrt{(-4)^2 + 4^2}$

$= 5.66$ km/hr

The vector $\begin{pmatrix} -4 \\ 4 \end{pmatrix}$ inclines at $45°$ to the x-axis,

so the direction of motion is $N\,45°\,W$ or bearing $315°$

3) *How far is the ship from port P at 1:30 pm ?*

When $t = \frac{3}{2}$, $r = \begin{pmatrix} 3 \\ 3 \end{pmatrix} + \frac{3}{2}\begin{pmatrix} 10 \\ 24 \end{pmatrix}$

$= \begin{pmatrix} 18 \\ 39 \end{pmatrix}$

Hence, distance from port P $= \sqrt{18^2 + 39^2}$

$= 43.0$ km

4) *Show that ship and the yacht collide and find the time and location of the collision.*

In order to collide, the ship and the yacht must be in the same place at the same time.

Let $$\begin{pmatrix} 3 \\ 3 \end{pmatrix} + t \begin{pmatrix} 10 \\ 24 \end{pmatrix} = \begin{pmatrix} 31 \\ 43 \end{pmatrix} + t \begin{pmatrix} -4 \\ 4 \end{pmatrix}$$

Using the x-coordinates :
$$3 + 10t = 31 - 4t$$
$$t = 2$$

Using the y-coordinates :
$$3 + 24t = 43 + 4t$$
$$t = 2$$

Since both equations give $t = 2$ this means that the ship and yacht will collide.

Time of collision = 2 pm

Location of collision = $\begin{pmatrix} 3 \\ 3 \end{pmatrix} + 2 \begin{pmatrix} 10 \\ 24 \end{pmatrix}$

$$= \begin{pmatrix} 23 \\ 51 \end{pmatrix}$$

= 23 km East and 51 km North from port P

Motion in three dimensions is very similar.

Consider a rocket flying along a path represented by

$$r = \begin{pmatrix} -1 \\ -2 \\ 5 \end{pmatrix} + t \begin{pmatrix} 3 \\ 6 \\ 2 \end{pmatrix}$$

where t is time in seconds and distance is measured in kilometres.

Initial location of the rocket $= (-1, -2, 5)$

This could represent a point 1 kilometre West, 2 kilometres South and 5 kilometres above a given origin $(0, 0, 0)$

Speed of the rocket $= \sqrt{3^2 + 6^2 + 2^2}$

$$= 7 \text{ km/sec}$$

$$= 25,200 \text{ km/hr}$$

It is often useful to remember that every point on the path of the rocket can be written in terms of the parameter t.

In this case the general point is $(-1 + 3t, \ -2 + 6t, \ 5 + 2t)$

Example

A model aircraft flies in a straight line such that its position vector
t minutes after 14: 00 is given by

$$\begin{pmatrix} x \\ y \\ z \end{pmatrix} = \begin{pmatrix} -15 \\ -17 \\ -25 \end{pmatrix} + t \begin{pmatrix} 3 \\ 6 \\ 5 \end{pmatrix}$$

A model rocket is launched at the same time   i.e. 14: 00

Its initial location is given by coordinates (3, 1, 2)

The rocket flies in a straight line and after 5 minutes it is
at the point (8, 21, 12).

All distances are measured in metres.

1) *Show that the position vector of the rocket after t minutes
can be written as*

$$\begin{pmatrix} x \\ y \\ z \end{pmatrix} = \begin{pmatrix} 3 \\ 1 \\ 2 \end{pmatrix} + t \begin{pmatrix} 1 \\ 4 \\ 2 \end{pmatrix}$$

The direction of the rocket is given by $\begin{pmatrix} 8 \\ 21 \\ 12 \end{pmatrix} - \begin{pmatrix} 3 \\ 1 \\ 2 \end{pmatrix}$ or $\begin{pmatrix} 5 \\ 20 \\ 10 \end{pmatrix}$

Dividing $\begin{pmatrix} 5 \\ 20 \\ 10 \end{pmatrix}$ by 5 gives the required parallel vector $\begin{pmatrix} 1 \\ 4 \\ 2 \end{pmatrix}$

Since the initial location is $(3, 1, 2)$ we have the required answer

$$\begin{pmatrix} x \\ y \\ z \end{pmatrix} = \begin{pmatrix} 3 \\ 1 \\ 2 \end{pmatrix} + t \begin{pmatrix} 1 \\ 4 \\ 2 \end{pmatrix}$$

2) *Show that the rocket hits the aircraft and find the time of impact.*

In order for the rocket to hit the aircraft there must be a unique value of t that satisfies the following :

$$\begin{pmatrix} -15 \\ -17 \\ -25 \end{pmatrix} + t \begin{pmatrix} 3 \\ 6 \\ 5 \end{pmatrix} = \begin{pmatrix} 3 \\ 1 \\ 2 \end{pmatrix} + t \begin{pmatrix} 1 \\ 4 \\ 2 \end{pmatrix}$$

Using the x-coordinates : $-15 + 3t = 3 + t$
$$t = 9$$

Using the y-coordinates : $-17 + 6t = 1 + 4t$
$$t = 9$$

Using the z-coordinates : $-25 + 5t = 2 + 2t$
$$t = 9$$

Hence the rocket will hit the aircraft at 14:09

(Note: if the value of t is not the same in all three equations then there will be no collision)

3) *Before hitting the aircraft, the rocket flies past a stationary balloon, B, located at coordinates (9, 8, 6).*
Calculate the closest distance between the rocket and the balloon.

A general point, R , on the straight line path of the rocket is

$$(3+t, \quad 1+4t, \quad 2+2t)$$

Hence $\overrightarrow{RB} = \begin{pmatrix} 9 \\ 8 \\ 6 \end{pmatrix} - \begin{pmatrix} 3+t \\ 1+4t \\ 2+2t \end{pmatrix}$

$$= \begin{pmatrix} 6-t \\ 7-4t \\ 4-2t \end{pmatrix}$$

The closest distance will occur when $\overrightarrow{RB}$ is perpendicular to the path of the rocket.

Using scalar product $\begin{pmatrix} 6-t \\ 7-4t \\ 4-2t \end{pmatrix} \cdot \begin{pmatrix} 1 \\ 4 \\ 2 \end{pmatrix} = 0$

$$6-t+28-16t+8-4t = 0$$

$$t = 2$$

Hence, at the closest approach $\overrightarrow{RB} = \begin{pmatrix} 4 \\ -1 \\ 0 \end{pmatrix}$

$$\left| \overrightarrow{RB} \right| = \sqrt{4^2 + (-1)^2 + 0^2}$$

$$= \sqrt{17}$$

$$= 4.12 \text{ m}$$

Exercise 2

1) Show that $\quad r = \begin{pmatrix} 2 \\ 11 \end{pmatrix} + \lambda \begin{pmatrix} 1 \\ 4 \end{pmatrix} \quad$ and $\quad y = 4x + 3$

represent the same line.

2) A line passes through the point (4, 5) and is parallel to the vector $\begin{pmatrix} 2 \\ 3 \end{pmatrix}$

   (a) write down the vector equation of the line

   (b) find the Cartesian equation of the line

3) Given $P = (3, 5)$ and $Q = (2, 8)$

    (a) find vector $\overrightarrow{PQ}$

    (b) find an equation of line PQ in the form $r = a + \lambda\, b$

    (c) show that the point $(1, 11)$ lies on line PQ

    (d) show that the point A with position vector $\begin{pmatrix} 6 \\ 2 \end{pmatrix}$

       does not lie on the line.

4)   A line passes through the point $(2, 5, 7)$ and is parallel to the vector $\begin{pmatrix} 2 \\ 3 \\ 4 \end{pmatrix}$.

    Write down a vector equation of the line.

5) Given $P = (1, 3, 5)$ and $Q = (5, 2, 8)$

    (a) find vector $\overrightarrow{PQ}$

    (b) find an equation of line PQ in the form $r = a + \lambda\, b$

    (c) show that the point $(9, 1, 11)$ lies on PQ .

    (d) show that the point A with position vector $\begin{pmatrix} 13 \\ 0 \\ 9 \end{pmatrix}$

       does not lie on the line.

6) (a) Find a vector equation of the line joining $(1, 7, 3)$ and $(3, 4, 6)$.

(b) Hence show that the points $(1, 7, 3)$, $(3, 4, 6)$ and $(5, 1, 9)$ are collinear.

7) Show that $r = \begin{pmatrix} 3 \\ 2 \end{pmatrix} + \mu \begin{pmatrix} 5 \\ 4 \end{pmatrix}$ and $5x + 4y = 8$ are perpendicular.

8) Show that $r = \begin{pmatrix} 1 \\ 5 \\ 8 \end{pmatrix} + t \begin{pmatrix} 1 \\ 2 \\ 5 \end{pmatrix}$ and $r = \begin{pmatrix} 2 \\ 3 \\ 4 \end{pmatrix} + k \begin{pmatrix} -3 \\ -11 \\ 5 \end{pmatrix}$

are perpendicular.

9) Find the point of intersection of

$$r = \begin{pmatrix} 1 \\ 4 \\ 2 \end{pmatrix} + t \begin{pmatrix} 1 \\ 2 \\ 5 \end{pmatrix} \quad \text{and} \quad r = \begin{pmatrix} 8 \\ -2 \\ 3 \end{pmatrix} + k \begin{pmatrix} -1 \\ 8 \\ 12 \end{pmatrix}$$

10) Show that $\quad r = \begin{pmatrix} 2 \\ 3 \\ 4 \end{pmatrix} + t\begin{pmatrix} 1 \\ 1 \\ 5 \end{pmatrix}$ and $\quad r = \begin{pmatrix} 8 \\ 1 \\ 2 \end{pmatrix} + k\begin{pmatrix} -4 \\ 4 \\ 1 \end{pmatrix}$

are skew.

11) Which of the following lines are parallel ?

(a) $\quad r = \begin{pmatrix} 1 \\ 2 \\ 3 \end{pmatrix} + \lambda\begin{pmatrix} 3 \\ 2 \\ -4 \end{pmatrix}$

(b) $\quad r = \begin{pmatrix} 2 \\ 5 \\ 8 \end{pmatrix} + \mu\begin{pmatrix} -3 \\ -2 \\ 4 \end{pmatrix}$

(c) $\quad r = \begin{pmatrix} 1 \\ 2 \\ 3 \end{pmatrix} + t\begin{pmatrix} -3 \\ 2 \\ 4 \end{pmatrix}$

12) Given quadrilateral $ABCD$ where

$A = (4, 3, 7)$ $B = (9, 5, 16)$ $C = (10, 5, 15)$ $D = (5, 3, 6)$

calculate the point of intersection of the diagonals AC and BD.

13) A ship moves such that its position vector relative to port P is given by

$$r = \begin{pmatrix} 5 \\ 3 \end{pmatrix} + t \begin{pmatrix} 9 \\ -12 \end{pmatrix}$$

where time t is measured in hours , distance is measured in kilometres and

$\begin{pmatrix} 1 \\ 0 \end{pmatrix}$ represents one kilometre East of port P.

$\begin{pmatrix} 0 \\ 1 \end{pmatrix}$ represents one kilometre North of port P.

(a) Calculate the speed of the ship.

(b) Calculate the bearing of the direction of motion of the ship.

A buoy is located at coordinates (16, 5)

(c) How far is the ship from the buoy after 80 minutes ?

(d) Calculate the closest distance between the ship and the buoy.

14) At 12 midnight a submarine is located at coordinates $(3, 6, -4)$.

This represents a position 3 km East and 6 km North from a port P, and a depth of 4 km below sea level.

The position vector of the submarine t hours after 12 midnight is given by

$$r = \begin{pmatrix} 3 \\ 6 \\ -4 \end{pmatrix} + t \begin{pmatrix} 6 \\ 3 \\ 1 \end{pmatrix}$$

(a) Calculate the speed of the submarine.

(b) At 12 midnight a torpedo is fired at the submarine. Its path is given by

$$r = \begin{pmatrix} 4 \\ 4 \\ -2 \end{pmatrix} + t \begin{pmatrix} 3 \\ 9 \\ -5 \end{pmatrix}$$

Show that the torpedo hits the submarine and calculate the time of impact.

(c) The torpedo does not explode and the submarine continues on the same course.
How far is the submarine from port P when it surfaces ?

Parametric and Cartesian line equations

$r = \begin{pmatrix} 1 \\ 2 \\ 3 \end{pmatrix} + \lambda \begin{pmatrix} 4 \\ 5 \\ 6 \end{pmatrix}$ is a vector equation of a straight line

$x = 1 + 4\lambda, \quad y = 2 + 5\lambda, \quad z = 3 + 6\lambda$ are the parametric equations of the same line.

$\dfrac{x-1}{4} = \dfrac{y-2}{5} = \dfrac{z-3}{6}$ is the Cartesian equation of the same line.

A method to quickly convert between the 3 different forms should be apparent from the above example.

However, if not, remember that $r = \begin{pmatrix} x \\ y \\ z \end{pmatrix}$ so we have $\begin{pmatrix} x \\ y \\ z \end{pmatrix} = \begin{pmatrix} 1 \\ 2 \\ 3 \end{pmatrix} + \lambda \begin{pmatrix} 4 \\ 5 \\ 6 \end{pmatrix}$

This gives the 3 parametric equations $x = 1 + 4\lambda, \quad y = 2 + 5\lambda, \quad z = 3 + 6\lambda$

Isolating λ gives $\dfrac{x-1}{4} = \lambda, \quad \dfrac{y-2}{5} = \lambda, \quad \dfrac{z-3}{6} = \lambda$

Hence $\dfrac{x-1}{4} = \dfrac{y-2}{5} = \dfrac{z-3}{6}$

44

Example

Write $\dfrac{2x-1}{4} = \dfrac{4-y}{5} = \dfrac{z+3}{6}$ as a vector equation

Step 1:

Rewrite the Cartesian equation as $\dfrac{x-\frac{1}{2}}{2} = \dfrac{y-4}{-5} = \dfrac{z+3}{6}$

Step 2:

Write down the answer $r = \begin{pmatrix} \frac{1}{2} \\ 4 \\ -3 \end{pmatrix} + \lambda \begin{pmatrix} 2 \\ -5 \\ 6 \end{pmatrix}$

Example

What is direction vector for the line $\dfrac{3x-2}{4} = \dfrac{9-2y}{5} = \dfrac{z+8}{6}$?

Step 1:

Rewrite the Cartesian equation as $\dfrac{x-\frac{2}{3}}{\frac{4}{3}} = \dfrac{y-\frac{9}{2}}{-\frac{5}{2}} = \dfrac{z+8}{6}$

Step 2:

The direction vector is given by $\begin{pmatrix} \frac{4}{3} \\ -\frac{5}{2} \\ 6 \end{pmatrix}$ or better still as $\begin{pmatrix} 8 \\ -15 \\ 36 \end{pmatrix}$

(Remember the vectors a and $6a$ have the same direction)

Example

Write down the coordinates of any point on $\quad \dfrac{3x-2}{4} = \dfrac{9-2y}{5} = \dfrac{z+8}{6}$

Step 1:

Rewrite the Cartesian equation as $\quad \dfrac{x-\frac{2}{3}}{\frac{4}{3}} = \dfrac{y-\frac{9}{2}}{-\frac{5}{2}} = \dfrac{z+8}{6}$

Step 2:

Write down the answer $\quad (\frac{2}{3}, \frac{9}{2}, -8)$

Note: here it is **incorrect** to multiply by 6
and give the answer as $(4, 18, -48)$.
The point $(4, 18, -48)$ does not lie on the line.

Exercise 3

1) Write down a vector equation of the following straight lines:

(a) $\dfrac{x-2}{2} = \dfrac{y+5}{-1} = \dfrac{z-4}{8}$

(b) $\dfrac{2x-6}{5} = \dfrac{8-y}{2} = \dfrac{4z-4}{3}$

(c) $\dfrac{x}{2} = \dfrac{4-y}{2} = z$

(d) $x = y = z$

(e) $\dfrac{x-8}{4} = \dfrac{y+2}{3}$, $z = 6$

(f) $x = 3+5\lambda,\quad y = 2-\lambda,\quad z = 4+3\lambda$

(g) $x = \lambda,\quad y = 6,\quad z = 4-5\lambda$

2) Calculate the point of intersection of the following straight lines :

$$\dfrac{x-1}{2} = \dfrac{y-3}{1} = \dfrac{z-1}{-1} \quad \text{and} \quad \dfrac{x-3}{3} = \dfrac{y-1}{3} = \dfrac{z+7}{2}$$

3) Which of the following lines are parallel ?

(a) $\dfrac{2x-1}{4} = \dfrac{8-y}{2} = \dfrac{z-5}{3}$

(b) $\dfrac{3x+2}{4} = \dfrac{y+1}{2} = \dfrac{z-9}{3}$

(c) $r = \begin{pmatrix} 1 \\ 2 \\ 3 \end{pmatrix} + \lambda \begin{pmatrix} 2 \\ -2 \\ 3 \end{pmatrix}$

(d) $x = 3+4\lambda, \quad y = 2+6\lambda, \quad z = 4+9\lambda$

4) Show that $\dfrac{2x-1}{6} = \dfrac{y+5}{2} = \dfrac{3-z}{5}$ and $\dfrac{x-1}{2} = \dfrac{y+6}{7} = \dfrac{2z+1}{8}$

are perpendicular.

5) Calculate the angle between the following pairs of straight lines:

(a) $r = \begin{pmatrix} 1 \\ 5 \\ 0 \end{pmatrix} + \lambda \begin{pmatrix} 2 \\ 1 \\ 3 \end{pmatrix}$ and $r = \begin{pmatrix} 7 \\ 2 \\ 3 \end{pmatrix} + \lambda \begin{pmatrix} 3 \\ 6 \\ 1 \end{pmatrix}$

(b) $\dfrac{3x+2}{6} = \dfrac{y+1}{2} = \dfrac{8-z}{3}$ and $\dfrac{x-2}{3} = \dfrac{3y+1}{9} = \dfrac{2-z}{2}$

(c) $x = y = z$ and $x = 3+5\lambda, \quad y = 2-\lambda, \quad z = 4+3\lambda$

6) Given point $P = (-2,9,5)$ and line $r = \begin{pmatrix} 4 \\ 1 \\ 3 \end{pmatrix} + t \begin{pmatrix} 1 \\ 2 \\ 1 \end{pmatrix}$

you are told that N is the foot of the perpendicular from P to the line.

(a) Write down the coordinates of N in terms of t.

(b) Hence find $\overrightarrow{PN}$ in terms of t.

(c) Using scalar product , calculate the value of t and hence find the actual coordinates of N.

(d) Hence calculate the shortest distance from P to the line.

Vector Product

The scalar product of two vectors will give an answer that is a scalar.

The vector product of two vectors will give an answer that is a vector.

The vector product is also known as the cross product.

The vector product of two vectors a and b is defined as

$$a \times b = |a||b| \sin \theta \ \hat{n}$$

where $\hat{n}$ is a unit vector perpendicular to both a and b

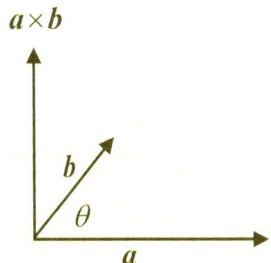

$|a||b| \sin \theta$ is the magnitude of the vector product.

The direction of $a \times b$ is found as follows:

> Imagine a screw turning in a circular motion from a to b; the direction in which the screw moves will be the direction of $a \times b$

Hence $a \times b = -b \times a$

Example

The angle between $\quad a = \begin{pmatrix} 1 \\ \sqrt{3} \\ 0 \end{pmatrix} \quad$ and $\quad b = \begin{pmatrix} 4\sqrt{3} \\ 4 \\ 0 \end{pmatrix} \quad$ is $30°$

Find $a \times b$

$$a \times b = |a||b| \sin\theta \ \hat{n}$$
$$= (2)(8) \sin 30° \ \hat{n}$$

$$= -8k \quad \text{since } k \text{ is a unit vector perpendicular} \\ \text{to both } i \text{ and } j$$

Note: the answer is negative due to the clockwise rotation of the screw when moving from a to b

This answer can also be found using the following method:

Write i, j, k and the two vectors as a matrix determinant (see Appendix 1)

$$a \times b = \begin{vmatrix} i & j & k \\ 1 & \sqrt{3} & 0 \\ 4\sqrt{3} & 4 & 0 \end{vmatrix}$$

$$= i \begin{vmatrix} \sqrt{3} & 0 \\ 4 & 0 \end{vmatrix} - j \begin{vmatrix} 1 & 0 \\ 4\sqrt{3} & 0 \end{vmatrix} + k \begin{vmatrix} 1 & \sqrt{3} \\ 4\sqrt{3} & 4 \end{vmatrix}$$

$$= i(0) - j(0) + k(4-12)$$

$$= -8k$$

Example

Given $a = \begin{pmatrix} 1 \\ 2 \\ 3 \end{pmatrix}$ and $b = \begin{pmatrix} 4 \\ 5 \\ 6 \end{pmatrix}$ calculate $a \times b$

Hence show that $\begin{pmatrix} 1 \\ -2 \\ 1 \end{pmatrix}$ is perpendicular to both a and b.

$$\begin{vmatrix} i & j & k \\ 1 & 2 & 3 \\ 4 & 5 & 6 \end{vmatrix} = i \begin{vmatrix} 2 & 3 \\ 5 & 6 \end{vmatrix} - j \begin{vmatrix} 1 & 3 \\ 4 & 6 \end{vmatrix} + k \begin{vmatrix} 1 & 2 \\ 4 & 5 \end{vmatrix}$$

$$= -3i + 6j - 3k$$

This vector is perpendicular to both a and b.

Since $\begin{pmatrix} -3 \\ 6 \\ -3 \end{pmatrix} = -3 \begin{pmatrix} 1 \\ -2 \\ 1 \end{pmatrix}$ we have $\begin{pmatrix} -3 \\ 6 \\ -3 \end{pmatrix}$ parallel to $\begin{pmatrix} 1 \\ -2 \\ 1 \end{pmatrix}$

but in the reverse direction.

Hence $\begin{pmatrix} 1 \\ -2 \\ 1 \end{pmatrix}$ is also perpendicular to both a and b.

Properties of Vector Product

1) Vector product is **not** commutative

$$a \times b = -b \times a \qquad \text{(see Page 49)}$$

2) Vector product is distributive over addition

$$a \times (b+c) = a \times b + a \times c$$

3) $(ka) \times b = k(a \times b)$ where k is a constant

4) $a \times a = 0$

since $|a||a|\sin 0 = 0$

Area of a triangle

The magnitude of $a \times b$ is $|a||b|\sin\theta$

This is the area formula for parallelogram $OACB$ shown below

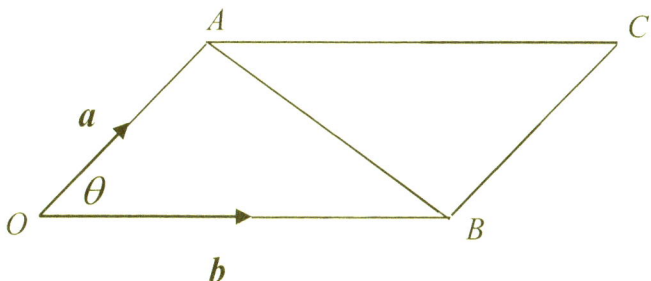

Hence the area of triangle OAB is $\dfrac{1}{2}|a||b|\sin\theta$

Example

Calculate the area of the triangle with sides

$$a = \begin{pmatrix} 1 \\ 2 \\ 3 \end{pmatrix} \quad \text{and} \quad b = \begin{pmatrix} 4 \\ 5 \\ 6 \end{pmatrix}$$

From the previous example $\quad a \times b = \begin{pmatrix} -3 \\ 6 \\ -3 \end{pmatrix}$

$$|a \times b| = \sqrt{9 + 36 + 9}$$
$$= \sqrt{54}$$

Hence $\quad$ area $= \dfrac{1}{2}\sqrt{54}$

$$= 3.67$$

Exercise 4

1) Calculate $a \times b$

(i) $a = \begin{pmatrix} 3 \\ -2 \\ 5 \end{pmatrix}$ and $b = \begin{pmatrix} 2 \\ 1 \\ -5 \end{pmatrix}$

(ii) $a = 8i + 6j - k$ and $b = i + 2j - 4k$

2) Find a vector perpendicular to both $\begin{pmatrix} 1 \\ -2 \\ 4 \end{pmatrix}$ and $\begin{pmatrix} -5 \\ 2 \\ 1 \end{pmatrix}$

3) A parallelogram has adjacent sides represented by $a = 2i + 3j - k$ and $b = 3i + 2j - k$. Calculate the area of the parallelogram.

4) A triangle has adjacent sides represented by $a = i + 2j - k$ and $b = 5i + j - 2k$. Calculate the area of the triangle.

5) Given $a = \begin{pmatrix} 2 \\ 5 \\ 8 \end{pmatrix}$ and $b = \begin{pmatrix} 6 \\ -1 \\ 2 \end{pmatrix}$

(a) Calculate   (i) $a \times b$ (ii) $|a \times b|$

                            (iii) $|a|$ (iv) $|b|$

(b) Hence find the angle between vectors a and b.

(c) Confirm your answer by using scalar product

Plane Equations

$3x + 4y + 5z = 25$ is the Cartesian equation of a plane.

$$\mathbf{r} \cdot \begin{pmatrix} 3 \\ 4 \\ 5 \end{pmatrix} = 25 \quad \text{is a vector equation of the same plane (in } \textit{normal} \text{ form)}$$

$$\mathbf{r} = \begin{pmatrix} 2 \\ 1 \\ 3 \end{pmatrix} + \lambda \begin{pmatrix} 1 \\ 3 \\ -3 \end{pmatrix} + \mu \begin{pmatrix} 7 \\ 1 \\ -5 \end{pmatrix} \quad \text{is another vector equation of the same plane.}$$

Consider $\mathbf{r} \cdot \begin{pmatrix} 3 \\ 4 \\ 5 \end{pmatrix} = 25$

R is the general point (x, y, z) that lies on the plane

Hence $\overrightarrow{OR} = \mathbf{r} = \begin{pmatrix} x \\ y \\ z \end{pmatrix}$

Therefore $\quad \mathbf{r} \cdot \begin{pmatrix} 3 \\ 4 \\ 5 \end{pmatrix} = 25 \quad$ can be written as $\quad \begin{pmatrix} x \\ y \\ z \end{pmatrix} \cdot \begin{pmatrix} 3 \\ 4 \\ 5 \end{pmatrix} = 25$

or $\quad 3x + 4y + 5z = 25$

In this format $\begin{pmatrix} 3 \\ 4 \\ 5 \end{pmatrix}$ is a vector *normal* to the plane

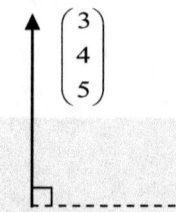

Finding a vector equation of a plane in *normal* form

A plane is uniquely defined if we know its orientation in space **and** its location in space.
The orientation is defined by a vector n normal to the plane.
The location is defined by a point on the plane

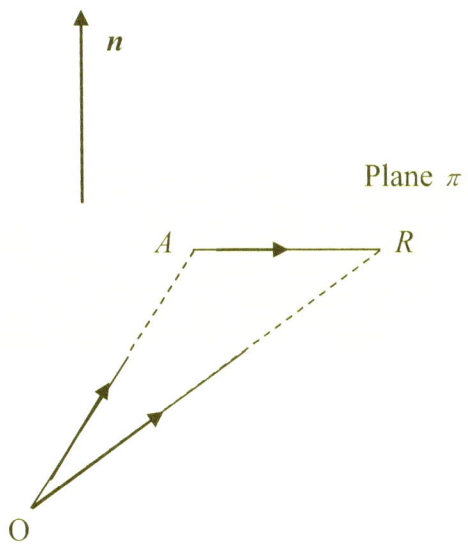

In the above diagram n is a vector normal to the plane π
and A is a fixed point on the the plane.
R is the general point (x, y, z) on the plane π

We know that $\overline{AR}$ is perpendicular to n

This means that $\overline{AR} \cdot n = 0$

Hence $(r - a) \cdot n = 0$

$r \cdot n - a \cdot n = 0$

$r \cdot n = a \cdot n$

Example

Find the equation of the plane that contains the point $(6, 1, 2)$

and is perpendicular to the vector $\begin{pmatrix} 3 \\ 4 \\ 5 \end{pmatrix}$.

Hence find the Cartesian equation of the plane

Step 1: $\quad\quad\quad r \bullet n = a \bullet n$

$$r \bullet \begin{pmatrix} 3 \\ 4 \\ 5 \end{pmatrix} = \begin{pmatrix} 6 \\ 1 \\ 2 \end{pmatrix} \bullet \begin{pmatrix} 3 \\ 4 \\ 5 \end{pmatrix}$$

$$= 18 + 4 + 10$$

$$= 32$$

Hence $\quad r \bullet \begin{pmatrix} 3 \\ 4 \\ 5 \end{pmatrix} = 32 \quad$ is the vector equation of the plane

in normal form.

Step 2: Convert to a Cartesian equation

$$r \bullet \begin{pmatrix} 3 \\ 4 \\ 5 \end{pmatrix} = 32$$

$$\begin{pmatrix} x \\ y \\ z \end{pmatrix} \bullet \begin{pmatrix} 3 \\ 4 \\ 5 \end{pmatrix} = 32$$

$$3x + 4y + 5z = 32$$

Example

Plane π is perpendicular to, and bisects, the line segment $\overline{PQ}$.
If $P = (2, 7, 4)$ and $Q = (8, 5, 12)$ find the Cartesian equation of plane π.

Step 1: Find vector $\overrightarrow{PQ}$

$$\overrightarrow{PQ} = \begin{pmatrix} 8 \\ 5 \\ 12 \end{pmatrix} - \begin{pmatrix} 2 \\ 7 \\ 4 \end{pmatrix} = \begin{pmatrix} 6 \\ -2 \\ 8 \end{pmatrix}$$

Step 2: Find mid-point of $\overline{PQ}$.

This point lies on the plane π.

$$\text{Mid-point of } \overline{PQ} = \left(\frac{2+8}{2}, \frac{7+5}{2}, \frac{4+12}{2} \right)$$

$$= (5, 6, 8)$$

Step 3: Use $\boldsymbol{r \cdot n = a \cdot n}$

$$\boldsymbol{r} \cdot \begin{pmatrix} 6 \\ -2 \\ 8 \end{pmatrix} = \begin{pmatrix} 5 \\ 6 \\ 8 \end{pmatrix} \cdot \begin{pmatrix} 6 \\ -2 \\ 8 \end{pmatrix}$$

$$\boldsymbol{r} \cdot \begin{pmatrix} 6 \\ -2 \\ 8 \end{pmatrix} = 30 - 12 + 64$$

$$\boldsymbol{r} \cdot \begin{pmatrix} 6 \\ -2 \\ 8 \end{pmatrix} = 82$$

Hence $\quad 6x - 2y + 8z = 82 \quad$ or $\quad 3x - y + 4z = 41$

Example

Show that the plane $r = \begin{pmatrix} 2 \\ 1 \\ 3 \end{pmatrix} + \lambda \begin{pmatrix} 1 \\ 3 \\ -3 \end{pmatrix} + \mu \begin{pmatrix} 7 \\ 1 \\ -5 \end{pmatrix}$ has Cartesian

equation $3x + 4y + 5z = 25$.

In this format $\begin{pmatrix} 1 \\ 3 \\ -3 \end{pmatrix}$ and $\begin{pmatrix} 7 \\ 1 \\ -5 \end{pmatrix}$ are vectors parallel to the given plane

and $\begin{pmatrix} 2 \\ 1 \\ 3 \end{pmatrix}$ is the position vector of the point $(2, 1, 3)$ on the plane.

We need to find a vector n perpendicular to the plane. In other words

we need to find a vector perpendicular to both $\begin{pmatrix} 1 \\ 3 \\ -3 \end{pmatrix}$ and $\begin{pmatrix} 7 \\ 1 \\ -5 \end{pmatrix}$

This is done by using vector product

$$\begin{pmatrix} 1 \\ 3 \\ -3 \end{pmatrix} \times \begin{pmatrix} 7 \\ 1 \\ -5 \end{pmatrix} = -12i - 16j - 20k$$

Hence $\begin{pmatrix} -12 \\ -16 \\ -20 \end{pmatrix}$ or $\begin{pmatrix} 3 \\ 4 \\ 5 \end{pmatrix}$ is the required vector

Using $r \bullet n = a \bullet n$

$$r \bullet \begin{pmatrix} 3 \\ 4 \\ 5 \end{pmatrix} = \begin{pmatrix} 2 \\ 1 \\ 3 \end{pmatrix} \bullet \begin{pmatrix} 3 \\ 4 \\ 5 \end{pmatrix} = 25$$

Hence $3x + 4y + 5z = 25$

Exercise 5

1) A plane contains the point $(5, 2, 1)$ and is parallel to the vectors
   $2i + 4j + 5k$ and $3i + 6j + k$

   (a) Find the vector equation of the plane in the form $r \cdot n = k$
       where $k \in Z$

   (b) Hence write down the Cartesian equation of the plane.

2) A plane contains the points $(3, 7, 1)$ and $(4, 9, 2)$ and is parallel to the
   vector $3i + 7j + k$. Find the Cartesian equation of the plane.

3) A plane contains the line $r = \begin{pmatrix} 4 \\ 1 \\ 3 \end{pmatrix} + \lambda \begin{pmatrix} 1 \\ 2 \\ 1 \end{pmatrix}$ and is parallel to $\begin{pmatrix} 2 \\ -5 \\ 8 \end{pmatrix}$.

   Find the Cartesian equation of the plane.

4) A plane contains the points $A(1, 5, 0)$, $B(2, 6, 1)$ and $C(7, 8, 4)$

   (a) Find the Cartesian equation of the plane.

   (b) Hence show that $A(1, 5, 0)$, $B(2, 6, 1)$, $C(7, 8, 4)$ and
       $D(5, 6, 2)$ are co-planar.

5) Find the Cartesian equation of the plane $r = \begin{pmatrix} 1 \\ 1 \\ 5 \end{pmatrix} + \lambda \begin{pmatrix} 2 \\ 1 \\ -2 \end{pmatrix} + \mu \begin{pmatrix} 3 \\ 1 \\ -4 \end{pmatrix}$

6) A plane π is parallel to $3x + 4y + 5z = 8$ and contains the point $(1, 6, 8)$. Find the Cartesian equation of plane π.

7) A plane is parallel to lines $r = \begin{pmatrix} 5 \\ 4 \\ 3 \end{pmatrix} + \lambda \begin{pmatrix} 3 \\ 2 \\ 1 \end{pmatrix}$ and $r = \begin{pmatrix} 1 \\ 6 \\ 2 \end{pmatrix} + \mu \begin{pmatrix} 1 \\ 4 \\ 1 \end{pmatrix}$

If the plane passes through the origin find its Cartesian equation.

8) Show that the line $r = \begin{pmatrix} 5 \\ 4 \\ 3 \end{pmatrix} + t \begin{pmatrix} 3 \\ 2 \\ 1 \end{pmatrix}$ is parallel to the plane

$2x - y - 4z = 8$

Show also that the given line does **not** lie in the plane $2x - y - 4z = 8$

9) A plane is perpendicular to $r = \begin{pmatrix} 5 \\ 6 \\ 2 \end{pmatrix} + t \begin{pmatrix} 3 \\ 4 \\ 1 \end{pmatrix}$ and passes through the

point $(3, 8, 2)$. Find the Cartesian equation of the plane.

10) Find the equation of the line perpendicular to $4x + 5y - 2z = 7$ given that the line passes through the point $(1, 2, 1)$

Intersection of a line and a plane

There are 3 possible situations:

1) the line cuts the plane at a single point

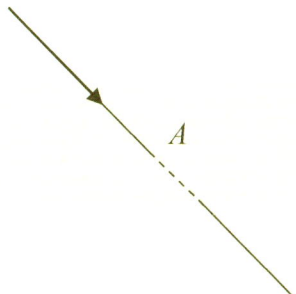

2) the line is parallel to the plane and does not lie in the plane
(in other words: no intersection; no solution)

3) the line is parallel to the plane and lies in the plane
(in other words: infinite solutions; any point that lies on the line)

Example

Find the point of intersection of the line $r = \begin{pmatrix} 1 \\ -3 \\ 3 \end{pmatrix} + \lambda \begin{pmatrix} 1 \\ 2 \\ 1 \end{pmatrix}$

and the plane $x + 2y + 3z = 20$

Step 1: Write the line equation in parametric form

$$\begin{aligned} x &= 1 + \lambda \\ y &= -3 + 2\lambda \\ z &= 3 + \lambda \end{aligned}$$

Step 2: Substitute into the plane equation and solve to find λ

$$(1 + \lambda) + 2(-3 + 2\lambda) + 3(3 + \lambda) = 20$$

$$1 + \lambda - 6 + 4\lambda + 9 + 3\lambda = 20$$

$$8\lambda + 4 = 20$$

$$\lambda = 2$$

Step 3: Substitute λ into the line equation to find the answer

$$r = \begin{pmatrix} 1 \\ -3 \\ 3 \end{pmatrix} + 2 \begin{pmatrix} 1 \\ 2 \\ 1 \end{pmatrix}$$

$$= \begin{pmatrix} 3 \\ 1 \\ 5 \end{pmatrix}$$

Hence the point of intersection is (3, 1, 5)

Example

Show that $r = \begin{pmatrix} 5 \\ 6 \\ 2 \end{pmatrix} + t \begin{pmatrix} 3 \\ 4 \\ 1 \end{pmatrix}$ does not intersect the plane $2x - 3y + 6z = 20$

Step 1: Show that the line and plane are parallel :

The direction of the line is given by $\begin{pmatrix} 3 \\ 4 \\ 1 \end{pmatrix}$

The normal to the plane has direction given by $\begin{pmatrix} 2 \\ -3 \\ 6 \end{pmatrix}$

$$\begin{pmatrix} 3 \\ 4 \\ 1 \end{pmatrix} \cdot \begin{pmatrix} 2 \\ -3 \\ 6 \end{pmatrix} = 6 - 12 + 6$$

$= 0 \qquad \Rightarrow$ line perpendicular to normal

$\Rightarrow$ line parallel to plane

Step 2: Show that a point on the line does not lie on the plane :

Substitute $(5, 6, 2)$ into the plane equation $2x - 3y + 6z = 20$

$10 - 18 + 12 \neq 20$

Since $(5, 6, 2)$ does not satisfy the plane equation, the line cannot lie in the plane.
Hence the line does not intersect the plane.

Example

Show that $r = \begin{pmatrix} 1 \\ 4 \\ 5 \end{pmatrix} + t \begin{pmatrix} 3 \\ 4 \\ 1 \end{pmatrix}$ lies in the plane $2x - 3y + 6z = 20$

Step 1: Show that the line and plane are parallel :

The direction of the line is given by $\begin{pmatrix} 3 \\ 4 \\ 1 \end{pmatrix}$

The normal to the plane has direction given by $\begin{pmatrix} 2 \\ -3 \\ 6 \end{pmatrix}$

$$\begin{pmatrix} 3 \\ 4 \\ 1 \end{pmatrix} \cdot \begin{pmatrix} 2 \\ -3 \\ 6 \end{pmatrix} = 6 - 12 + 6$$

$= 0 \quad \Rightarrow$ line perpendicular to normal

$\Rightarrow$ line parallel to plane

Step 2: Show that a point on the line also lies on the plane :

Substitute $(1, 4, 5)$ into the plane equation $2x - 3y + 6z = 20$

giving $2 - 12 + 30 = 20$

Since $(1, 4, 5)$ satisfies the plane equation, the line lies in the plane.

Angle between a line and a plane

This is usually the acute angle θ shown below:

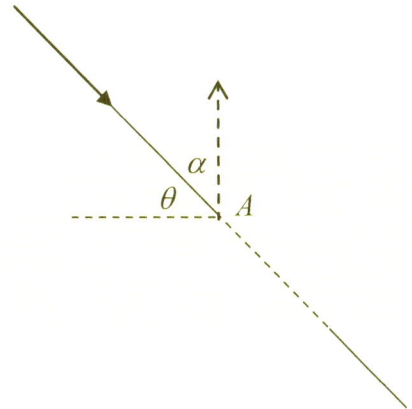

Step 1: Find the angle α between the line and the normal to the plane using scalar product.

Step 2: $\theta = 90° - \alpha$

The direction of the line is represented by vector b in the line equation
$r = a + \lambda b$

The normal to the plane is represented by vector n in the plane equation
$r \cdot n = k$

Example

Find the angle between the line $\quad r = \begin{pmatrix} 1 \\ 4 \\ 5 \end{pmatrix} + \lambda \begin{pmatrix} 3 \\ 4 \\ 1 \end{pmatrix} \quad$ and the plane

$2x + 3y + 5z = 8$

Step 1: Find the angle between the line and the normal to the plane.

The direction of the line is given by $\begin{pmatrix} 3 \\ 4 \\ 1 \end{pmatrix}$

The direction of the normal to the plane is given by $\begin{pmatrix} 2 \\ 3 \\ 5 \end{pmatrix}$

Using scalar product

$$\begin{pmatrix} 3 \\ 4 \\ 1 \end{pmatrix} \cdot \begin{pmatrix} 2 \\ 3 \\ 5 \end{pmatrix} \quad = \quad \sqrt{3^2 + 4^2 + 1^2} \ \sqrt{2^2 + 3^2 + 5^2} \ \cos\alpha$$

$$6 + 12 + 5 = \sqrt{26} \ \sqrt{38} \ \cos\alpha$$

$$\cos\alpha = \frac{23}{\sqrt{26}\sqrt{38}}$$

$$\alpha = 43.0°$$

Step 2: Calculate θ

$$\theta = 90° - \alpha$$
$$= 47.0°$$

Angle between two planes

The angle between two planes has the same value as the angle between their respective normals .

Example

Find the angle between the planes $5x + 2y + 3z = 12$ and $4x + y + 6z = 9$

The directions of the normals are $\begin{pmatrix} 5 \\ 2 \\ 3 \end{pmatrix}$ and $\begin{pmatrix} 4 \\ 1 \\ 6 \end{pmatrix}$

Using scalar product

$$\begin{pmatrix} 5 \\ 2 \\ 3 \end{pmatrix} \cdot \begin{pmatrix} 4 \\ 1 \\ 6 \end{pmatrix} = \sqrt{5^2 + 2^2 + 3^2}\ \sqrt{4^2 + 1^2 + 6^2}\ \cos\theta$$

$$20 + 2 + 18 = \sqrt{38}\ \sqrt{53}\ \cos\theta$$

$$\cos\theta = \frac{40}{\sqrt{38}\sqrt{53}}$$

$$\theta = 27.0°$$

Exercise 6

1) Given line $r = \begin{pmatrix} -5 \\ 3 \\ 2 \end{pmatrix} + \lambda \begin{pmatrix} 2 \\ 0 \\ 1 \end{pmatrix}$ and plane $3x - y + 2z = 10$

   (a) find the coordinates of the point of intersection.

   (b) calculate the angle between the line and the plane

2) Given line $x = \dfrac{y-7}{-2} = \dfrac{z-1}{3}$ and plane $x + y + z = 12$

   (a) find the coordinates of the point of intersection.

   (b) calculate the angle between the line and the plane

3) Calculate the angle between the following two planes:

   $\pi_1 :$ $2x - 5y + 4z = 2$

   $\pi_2 :$ $3x + y + 6z = 8$

4) Calculate the point of intersection of the plane $r \cdot \begin{pmatrix} 3 \\ 1 \\ 2 \end{pmatrix} = 15$

   and the line $r = \begin{pmatrix} 5 \\ 3 \\ 6 \end{pmatrix} + \lambda \begin{pmatrix} 4 \\ -5 \\ 4 \end{pmatrix}$

5) Given $\pi_1 :$ $2x - 3y + 5z = 6$
            $\pi_2 :$ $4x + 6y + 2z = 13$

   show that π_1 and π_2 are perpendicular

Intersection of two planes

There are three possible situations:

1) the planes are parallel and distinct

2) the planes are parallel and coincide

3) the planes intersect to form a straight line

First, check to see if the planes are parallel. To do this, simply check to see if the two planes have parallel normals.

If the planes are parallel and distinct then there is no solution.
If the planes are parallel and coincide then there are infinite solutions
(any point on the plane)

If the planes are not parallel then they will meet to form a straight line.

Example

Find the equation of the line of intersection of the two planes

π_1 : $3x + 10y + z = 26$

π_2 : $x + 6y + z = 16$

Step 1:   Solve the two plane equations simultaneously, eliminating one of the variables

π_1 : $3x + 10y + z = 26$

π_2 : $x + 6y + z = 16$

Subtracting the two equations gives $2x + 4y = 10$

Solving for x: $x = 5 - 2y$

Step 2:   Repeat the above process but this time eliminate y

Multiply π_1 by 3 ;   multiply π_2 by 5

π_1 : $9x + 30y + 3z = 78$

π_2 : $5x + 30y + 5z = 80$

Subtracting the two equations: $4x - 2z = -2$

Solving for x : $x = \dfrac{z-1}{2}$

Step 3:   Write down the line equation

$$x = 5 - 2y = \frac{z-1}{2}$$

Intersection of three planes

There are seven possible situations:

1) the three planes intersect to form a point

2) the three planes intersect to form a straight line

3)   the three planes are parallel and coincide

4)   the three planes are parallel and distinct

5)   two of the planes are parallel and coincide , and the third plane
     is parallel and distinct

6)   two of the planes are parallel and distinct and the third plane
     cuts across the other two planes

7) the line of intersecton of two of the planes is parallel to the third plane

Note: In the first three situations an answer is possible.

In the last four situations there is no solution;
however you may be required to give a geometrical explanation.

The first check is always to see if the planes are parallel.
(check for parallel normals).

Example

Find the point of intersection of the given three planes:

$$\pi_1 : \qquad 2x + y + z = 7$$

$$\pi_2 : \qquad x - y + z = 2$$

$$\pi_3 : \qquad -x + 2y + z = 6$$

The solution can be found either by using simultaneous equations or by using matrices.

Method 1:

Subtract π_2 from π_1 giving $\quad x + 2y = 5$

Subtract π_3 from π_1 giving $\quad 3x - y = 1$

Solve simultaneously to give $\quad x = 1$ and $y = 2$

Substitute into π_1 to find $\quad z = 3$

Hence the point of intersection is $\quad (1, 2, 3)$

Method 2   (optional : see Appendix II   and Appendix III)

Write the three plane equations as a matrix product:

$$\begin{pmatrix} 2 & 1 & 1 \\ 1 & -1 & 1 \\ -1 & 2 & 1 \end{pmatrix} \begin{pmatrix} x \\ y \\ z \end{pmatrix} = \begin{pmatrix} 7 \\ 2 \\ 6 \end{pmatrix}$$

Solve with a GDC (eg. TI 84 plus)

$$\begin{pmatrix} x \\ y \\ z \end{pmatrix} = \begin{pmatrix} 2 & 1 & 1 \\ 1 & -1 & 1 \\ -1 & 2 & 1 \end{pmatrix}^{-1} \begin{pmatrix} 7 \\ 2 \\ 6 \end{pmatrix}$$

$$= \begin{pmatrix} 1 \\ 2 \\ 3 \end{pmatrix}$$

Hence the point of intersection is   (1, 2, 3)

Example

Give a geometrical description of the following three planes:

$$\pi_1 : \qquad x - y - z = 0$$

$$\pi_2: \qquad 2x - 3y - z = 3$$

$$\pi_3: \qquad x - 3y + z = 6$$

The planes are not parallel (the normals all have different directions) so the only possible answers are:

(a) the three planes intersect to form a single point

(b) the three planes intersect to form a straight line

(c) the line of intersecton of two of the planes is parallel to the third plane.

Subtract π_1 from π_2 giving $\quad x - 2y = 3$

Add π_2 to π_3 giving $\quad 3x - 6y = 9$

These two results are consistent so we have $\quad x = 2y + 3$

Subtract π_3 from π_2 giving $\quad x - 2z = -3$

Subtract π_2 from three times π_1 giving $\quad x - 2z = -3$

Again these two results are consistent, so we have $\quad x = 2z - 3$

Hence the three planes intersect to form a straight line with equation

$$x = 2y + 3 = 2z - 3$$

Example

Give a geometrical description of the following three planes:

$$\pi_1 : \qquad x - y - z = 1$$

$$\pi_2 : \qquad 2x + 3y - z = 4$$

$$\pi_3 : \qquad x + 9y + z = 2$$

Again the planes are not parallel so the only possible solutions are:

(a) the three planes intersect to form a single point

(b) the three planes intersect to form a straight line

(c)   the line of intersecton of two of the planes is parallel to the third plane.

Subtract π_1 from π_2 giving $x + 4y = 3$

Add π_2 to π_3 giving $3x + 12y = 6$

This simplifies to $x + 4y = 2$

These two results are inconsistent so there is no solution and the answer must be that the line of intersecton of two of the planes is parallel to the third plane.

Example

Give a geometrical description of the following three planes:

$$\pi_1 : \quad 2x - 3y - z = 8$$

$$\pi_2 : \quad 4x - 6y - 2z = 5$$

$$\pi_3 : \quad x + 9y + z = 2$$

Planes π_1 and π_2 are parallel and distinct.

Plane π_3 cuts across both of these planes.

Exercise 7

1) Find the equation of the line of intersection of the two planes

(a) $\quad \pi_1 : x - y - z = 1 \qquad$ and $\qquad \pi_2 : 2x - y + z = 6$

(b) $\quad r \cdot \begin{pmatrix} 3 \\ 1 \\ 2 \end{pmatrix} = 8 \qquad$ and $\qquad r \cdot \begin{pmatrix} 1 \\ 1 \\ 1 \end{pmatrix} = 6$

2) Find the coordinates of the point of intersection of the following three planes:

π_1 : $x + y + 2z = 14$

π_2 : $3x - y - z = 3$

π_3 : $2x + y - z = 2$

3) Find the equation of the line of intersection of the following three planes :

π_1 : $3x + y + z = 4$

π_2 : $x + y - z = 2$

π_3 : $2x + 3y - 4z = 5$

4) (a)   Show that the following three planes do not intersect to form a line :

π_1 : $2x - y - z = 8$

π_2 : $x + y - z = 2$

π_3 : $7x + y - 5z = 8$

(b)   Give a brief geometrical explanation

5) Solve the following systems of equations.
(Each system represents three planes).
If there is no solution give a reason.

(a) π_1 : $2x + y + 3z = 8$

          π_2 : $2x + y + 3z = 10$

          π_3 : $x + 4y + z = 5$

(b) π_1 : $x + y + z = 8$

          π_2 : $2x + 2y + 2z = 16$

          π_3 : $-3x - 3y - 3z = -24$

(c) π_1 : $x + y + z = 8$

          π_2 : $2x + 2y + 2z = 10$

          π_3 : $-3x - 3y - 3z = -12$

(d) π_1 : $x + y + z = 4$

          π_2 : $x - y + z = 2$

          π_3 : $-x + 2y + z = 7$

6) Find the values of λ for which the following system of equations do **not** have a unique solution.

$$\lambda x + y + z = 10$$

$$x - 2y + z = 8$$

$$x - y + \lambda z = 1$$

7) Given the three planes

$\pi_1 :$ $2x + y + z = 6$

$\pi_2 :$ $x + y - z = 4$

$\pi_3 :$ $5x + 3y + z = k$ where k is a constant.

(a) Show that the three planes do **not** meet at a single point.

(b) (i)   Find the value of k for which the three planes intersect to form a straight line.

   (ii)   For this value of k, find the equation of the straight line.

Miscellaneous

1. The position vectors of points P and Q are $\begin{pmatrix} -1 \\ 2 \\ 5 \end{pmatrix}$ and $\begin{pmatrix} 2 \\ 3 \\ 6 \end{pmatrix}$

respectively. The origin is at O.

Find (a) $\overrightarrow{OP} \times \overrightarrow{OQ}$

(b) the area of the parallelogram with two sides $\overrightarrow{OP}$ and $\overrightarrow{OQ}$.

2. A triangle has its vertices at $A(-2, 5, 1)$, $B(4, 6, 2)$ and $C(5, 0, 3)$.

(a) Calculate $\overrightarrow{AB} \cdot \overrightarrow{AC}$

(b) Hence find angle BAC

3. Plane π contains the following lines L_1 and L_2

L_1 : $r = i - 2j + 3k + \lambda\,(i + 4j - k)$
L_2 : $r = 2i + 5j - k + \mu\,(i + j + 2k)$

(a) (i) Find a vector that is normal to the plane π

(ii) Hence find the equation of the plane π

(b) Calculate the angle between lines L_1 and L_2

4. Given that $a = i + 2j - k$, $b = 3i + j + 2k$ and $c = i - j + k$ are the position vectors of the points A, B and C respectively, calculate the area of triangle ABC.

5. Given that the points $P(2, 3, 8)$ and $Q(1, -3, 6)$ lie on the line l.

Find an equation of line l, giving your answer in

(a) parametric form. (b) Cartesian form

6. Given that $a = \begin{pmatrix} 1 \\ 3 \\ 6 \end{pmatrix}$, $b = \begin{pmatrix} 2 \\ 1 \\ 5 \end{pmatrix}$ and $c = \begin{pmatrix} -3 \\ 2 \\ 1 \end{pmatrix}$

find $(a \times b) \cdot c$.

7. The lines L_1 and L_2 have parametric equations

$$L_1: \quad x = 3 + \lambda, \quad y = 2 - \lambda, \quad z = 1 + 4\lambda$$

$$L_2: \quad x = 6 - \mu, \quad y = -2 + 2\mu, \quad z = 3 + 6\mu$$

(a) Calculate the angle between lines L_1 and L_2

(b) Find the coordinates of the point of intersection of L_1 and L_2.

8. The vector $n = i + j - 4k$ is normal to the plane π which passes through the point $(1, 5, -2)$.

    (a) Find the Cartesian equation of the plane π

    (b) Find p if the point $(p, p-4, p+2)$ lies on plane.

9. Line L intersects plane π at the point A

    $$L: \quad \frac{x-1}{2} = \frac{y+1}{1} = \frac{z-5}{3}$$

    $$\pi: \quad 2x + y + z = 22$$

    Find the coordinates of point A .

10. The plane $x - y + z = 7$ contains the line $\frac{x-2}{3} = \frac{y+1}{2} = \frac{z-4}{p}$.

    Find the value of p

11. Find the equation of the line of intersection of the two planes

    $$x + y + z = 5 \quad \text{and} \quad x - y + 3z = 3$$

12. The point P is the foot of the perpendicular from the point $(4, 2, 1)$ to the plane $x + 2y - z = 1$. Find the coordinates of P.

13. A plane π has equation $r \cdot \begin{pmatrix} 4 \\ -1 \\ 2 \end{pmatrix} = 15$ and a line l has

Cartesian equation $\dfrac{x-3}{2} = \dfrac{y-1}{2} = \dfrac{2-z}{3}$

Show that the line l lies in the plane π.

14. Given point A $(5, 8, 1)$ and plane $\pi: 2x + y - z = 5$

(a) Find the equation of the line perpendicular to the plane π that passes through the point A

(b) Let A' be the reflection of A in the plane π. Find the coordinates of the point A'.

15. Find the acute angle θ between the following planes π_1 and π_2 :

$$\pi_1 : \quad -3x + 2y - z = 5$$

$$\pi_2 : \quad 4x + y - 2z = 8$$

16. The points A, B, C have position vectors $i + j + k$, $3i + 2j + 3k$ and $2i + 4j - k$ respectively.

    (a)   (i)   Find $\overrightarrow{AB} \times \overrightarrow{BC}$

           (ii)   Find the cartesian equation of the plane containing points A, B, C

    (b)   (i)   Find the area of the triangle ABC

           (ii)   Hence, or otherwise, calculate the shortest distance from C to the line AB.

17. The points $P(4, 0, 1)$ and $Q(2, 2, 3)$ lie on the line L_1

    (a)   Find a set of parametric equations for line L_1

    (b)   The point A is on L_1 such that $\overrightarrow{OA}$ is perpendicular to L_1

           (i)   Find the coordinates of A.

           (ii)   Hence calculate the shortest distance from the origin O to the line PQ

           (iii)   Calculate the area of triangle OPQ.

18. Two planes π_1 and π_2 are represented by the equations

$$\pi_1 : \quad r = \begin{pmatrix} 1 \\ 4 \\ 2 \end{pmatrix} + \lambda \begin{pmatrix} 2 \\ 1 \\ -3 \end{pmatrix} + \mu \begin{pmatrix} 1 \\ 1 \\ -2 \end{pmatrix}$$

$$\pi_2 : \quad 3x - 4y + z = 4.$$

(a)   (i)   Find $\begin{pmatrix} 2 \\ 1 \\ -3 \end{pmatrix} \times \begin{pmatrix} 1 \\ 1 \\ -2 \end{pmatrix}$

(ii)   Show that the equation of π_1 can be written as
$x + y + z = 7$

(b)   Show that π_1 is perpendicular to π_2

(c)   Find the Cartesian equation of the line of intersection of π_1 and π_2

(d)   The point A has coordinates $(4, 6, 6)$
Let B be the foot of the perpendicular from A to the plane π_1

(i)   Find the coordinates of B

(ii)   Point A' is the reflection of point A in the plane π_1.
Find the coordinates of A'

19. Consider the points $A\,(2, 1, 1)$ and $B\,(4, 5, 2)$

(a) Calculate the cosine of the angle between $\overrightarrow{OA}$ and $\overrightarrow{OB}$, where O is the origin $(0, 0, 0)$

(b) Find a vector equation of the line L_1 which passes through A and B.

The line L_2 has equation $r = 10i + j + 4k + \mu\,(3i - 2j + k)$

where $\mu \in \mathbb{R}$

(c) Show that the lines L_1 and L_2 intersect and find the coordinates of their point of intersection.

20. (a) Find the coordinates of the point of intersection of the three planes

$$2x - y - z = 2$$

$$x + y + 3z = 10$$

$$5x - 2y + z = 15$$

(b) A fourth plane passes through the point of intersection. This plane has equation $x + 2y + 3z = k$. Find the value of k.

Answers

Exercise 1

1)   (a) $\begin{pmatrix} 2 \\ -2 \\ 7 \end{pmatrix}$ (b) $\sqrt{57}$

2)   (a) $20 \cdot 3°$ (b) $43 \cdot 3°$ (c) $126°$

4)   (a) (i) 3 (ii) $\sqrt{41}$ (b) $117 \cdot 94°$ (c) 8.49

6)   (a) $\frac{3}{5}i + \frac{4}{5}j$ (b) $\frac{1}{7}\begin{pmatrix} 2 \\ 3 \\ 6 \end{pmatrix}$ (c) $\frac{3}{\sqrt{11}}i + \frac{1}{\sqrt{11}}j + \frac{1}{\sqrt{11}}k$

7) $4i - 12j + 6k$

8)   (a) -6 (b) $\frac{8}{3}$

9) $k = -25$

Exercise 2

2) (a) $r = \begin{pmatrix} 4 \\ 5 \end{pmatrix} + \lambda \begin{pmatrix} 2 \\ 3 \end{pmatrix}$ (b) $y = \frac{3}{2}x - 1$

3) (a) $\begin{pmatrix} -1 \\ 3 \end{pmatrix}$ (b) $r = \begin{pmatrix} 3 \\ 5 \end{pmatrix} + \lambda \begin{pmatrix} -1 \\ 3 \end{pmatrix}$

4) $r = \begin{pmatrix} 2 \\ 5 \\ 7 \end{pmatrix} + \lambda \begin{pmatrix} 2 \\ 3 \\ 4 \end{pmatrix}$

5) (a) $\begin{pmatrix} 4 \\ -1 \\ 3 \end{pmatrix}$ (b) $r = \begin{pmatrix} 1 \\ 3 \\ 5 \end{pmatrix} + \lambda \begin{pmatrix} 4 \\ -1 \\ 3 \end{pmatrix}$

6) (a) $r = \begin{pmatrix} 1 \\ 7 \\ 3 \end{pmatrix} + \lambda \begin{pmatrix} 2 \\ -3 \\ 3 \end{pmatrix}$

9) (6, 14, 27)

11) (a) and (b)

12) (7, 4, 11)

13) (a) 15 km/hr   (b) $143°$ (c) 18.0 km   (d) 10 km

14) (a) 6.78 km/hr   (b) 12:20 am   (c) 32.4 km

Exercise 3

1)　(a)　$r = \begin{pmatrix} 2 \\ -5 \\ 4 \end{pmatrix} + \lambda \begin{pmatrix} 2 \\ -1 \\ 8 \end{pmatrix}$　(b)　$r = \begin{pmatrix} 3 \\ 8 \\ 1 \end{pmatrix} + \lambda \begin{pmatrix} 10 \\ -8 \\ 3 \end{pmatrix}$

(c)　$r = \begin{pmatrix} 0 \\ 4 \\ 0 \end{pmatrix} + \lambda \begin{pmatrix} 2 \\ -2 \\ 1 \end{pmatrix}$　(d)　$r = \lambda \begin{pmatrix} 1 \\ 1 \\ 1 \end{pmatrix}$

(e)　$r = \begin{pmatrix} 8 \\ -2 \\ 6 \end{pmatrix} + \lambda \begin{pmatrix} 4 \\ 3 \\ 0 \end{pmatrix}$　(f)　$r = \begin{pmatrix} 3 \\ 2 \\ 4 \end{pmatrix} + \lambda \begin{pmatrix} 5 \\ -1 \\ 3 \end{pmatrix}$

(g)　$r = \begin{pmatrix} 0 \\ 6 \\ 4 \end{pmatrix} + \lambda \begin{pmatrix} 1 \\ 0 \\ -5 \end{pmatrix}$

2)　$(9, 7, -3)$

3)　(a) and (c)　　(b) and (d)

5)　(a) 53.8　(b) 21.4　(c) 46.9

6)　(a)　$(4 + t, 1 + 2t, 3 + t)$　(b)　$\begin{pmatrix} 6 + t \\ -8 + 2t \\ -2 + t \end{pmatrix}$

(c) $t = 2$, $(6, 5, 5)$　(d) $4\sqrt{5}$

Exercise 4

1) (i) $\begin{pmatrix} 5 \\ 25 \\ 7 \end{pmatrix}$ (ii) $-22i + 31j + 10k$

2) $\begin{pmatrix} -10 \\ -21 \\ -8 \end{pmatrix}$

3) $\sqrt{27}$

4) $\dfrac{3\sqrt{11}}{2}$

5) (a) (i) $\begin{pmatrix} 18 \\ 44 \\ -32 \end{pmatrix}$ (ii) $\sqrt{3284}$ (iii) $\sqrt{93}$ (iv) $\sqrt{41}$

   (b) 68.1

Exercise 5

1)   (a) $r \cdot \begin{pmatrix} -2 \\ 1 \\ 0 \end{pmatrix} = -8$ (b) $-2x + y = -8$

2) $-5x + 2y + z = 0$

3) $21x - 6y - 9z = 51$

4)   (a) $x + 2y - 3z = 11$

5) $2x - 2y + z = 5$

6) $3x + 4y + 5z = 67$

7) $x + y - 5z = 0$

9) $3x + 4y + z = 43$

10) $r = \begin{pmatrix} 1 \\ 2 \\ 1 \end{pmatrix} + \lambda \begin{pmatrix} 4 \\ 5 \\ -2 \end{pmatrix}$

Exercise 6

1) (a) $(1, 3, 5)$ (b) $73.0°$

2) (a) $(2, 3, 7)$ (b) $18.0°$

3) $56.7°$

4) $(1, 8, 2)$

Exercise 7

1) (a) $x = \dfrac{2y+7}{3} = 5 - 2z$ (b) $x = y - 4 = \dfrac{2-z}{2}$

2) $(3, 1, 5)$

3) $x = \dfrac{3-y}{2} = 1 - z$

4) (b) Each plane is parallel to the line of intersection of the other two

5) (a) No solution: π_3 cuts across two parallel planes

    (b) Infinite solutions: any point on the plane $x + y + z = 8$

    (c) No solution: three parallel planes.

    (d) $(-1, 1, 4)$

6) $\lambda = \pm 1$

7) (b) (i) $k = 16$ (ii) $x = \dfrac{10 - 2y}{3} = 2 - 2z$

Miscellaneous Vectors

1. (a) $-3i + 16j - 7k$ (b) 17.7

2. (a) 39   (b) 44.2°

3. (a) (i) $3i - j - k$ (ii) $3x - y - z = 2$ (b) 73.2°

4. 5.02

5. (a) $x = 2 + \lambda$, $y = 3 + 6\lambda$, $z = 8 + 2\lambda$

    (b) $x - 2 = \dfrac{y-3}{6} = \dfrac{z-8}{2}$

6. -18

7. (a) 39.4°   (b) (5, 0, 9)

8. (a) $x + y - 4z = 14$ (b) -13

9. (5, 1, 11)

10. -1

11. $x = 6 - 2y = 4 - 2z$

12. (3, 0, 2)

14. (a) $r = \begin{pmatrix} 5 \\ 8 \\ 1 \end{pmatrix} + \lambda \begin{pmatrix} 2 \\ 1 \\ -1 \end{pmatrix}$ (b) $(-3, 4, 5)$

15. 62.2°

16. (a) (i) $-8i + 6j + 5k$ (ii) $-8x + 6y + 5z = 3$

(b) (i) 5.59 (ii) 3.73

17. (a) $x = 4 - \lambda, \quad y = \lambda, \quad z = 1 + \lambda$

(b) (i) $(3, 1, 2)$ (ii) $\sqrt{14}$ (iii) 6.48

18. (a) (i) $i + j + k$ (c) $x = \dfrac{5y-3}{2} = \dfrac{32-5z}{7}$

(d) (i) $(1, 3, 3)$ (ii) $(-2, 0, 0)$

19. (a) $\dfrac{\sqrt{30}}{6}$ (b) $r = \begin{pmatrix} 2 \\ 1 \\ 1 \end{pmatrix} + \lambda \begin{pmatrix} 2 \\ 4 \\ 1 \end{pmatrix}$ (c) $(4, 5, 2)$

20. (a) $(2, -1, 3)$ (b) $k = 9$

Appendix I

If $\quad A = \begin{pmatrix} a & b \\ c & d \end{pmatrix} \quad$ then the determinant of matrix A is given by

$$\begin{vmatrix} a & b \\ c & d \end{vmatrix} = ad - bc$$

If $\quad A = \begin{pmatrix} a & b & c \\ d & e & f \\ g & h & i \end{pmatrix} \quad$ then the determinant of matrix A is given by

$$\begin{vmatrix} a & b & c \\ d & e & f \\ g & h & i \end{vmatrix} = a \begin{vmatrix} e & f \\ h & i \end{vmatrix} - b \begin{vmatrix} d & f \\ g & i \end{vmatrix} + c \begin{vmatrix} d & e \\ g & h \end{vmatrix}$$

Appendix II

Finding the inverse of a 3×3 matrix on a TI 84 Plus :

Press 2nd x^{-1} to access MATRIX

Use the arrow keys to highlight EDIT

Press ENTER

Press 3 ENTER

Press 3 ENTER

A **three by three** matrix of should now be shown on the screen

Enter the required numbers.
For example: 5 ENTER, 1 ENTER, 3 ENTER, etc.
to display the matrix

$$\begin{pmatrix} 5 & 1 & 3 \\ 4 & 6 & 8 \\ 7 & 8 & 9 \end{pmatrix}$$

Press 2nd MODE to access QUIT

Press 2nd x^{-1} to access MATRIX again

Press ENTER

[A] should appear on the screen

Press x^{-1}

Press ENTER

The inverse matrix will now be displayed.

$$\begin{pmatrix} 0.167 & -0.25 & 0.167 \\ -0.333 & -0.4 & 0.467 \\ 0.167 & 0.55 & -0.433 \end{pmatrix}$$

As is the case here, the inverse matrix often does not consist of integers. If you cannot see all nine elements then use the arrow keys to scroll right.

It is often easier to convert decimals in a matrix to fractions. To do this, just press MATH ENTER ENTER

$$\begin{pmatrix} \frac{1}{6} & -\frac{1}{4} & \frac{1}{6} \\ -\frac{1}{3} & -\frac{2}{5} & \frac{7}{15} \\ \frac{1}{6} & \frac{11}{20} & -\frac{13}{30} \end{pmatrix}$$

Appendix III

Solving a matrix equation :

$$\text{Solve} \quad \begin{pmatrix} 5 & 1 & 3 \\ 4 & 6 & 8 \\ 7 & 8 & 9 \end{pmatrix} \begin{pmatrix} x \\ y \\ z \end{pmatrix} = \begin{pmatrix} 12 \\ 40 \\ 60 \end{pmatrix}$$

Enter $[A] = \begin{pmatrix} 5 & 1 & 3 \\ 4 & 6 & 8 \\ 7 & 8 & 9 \end{pmatrix}$ as before

Press 2nd MODE to access QUIT

Press 2nd x^{-1} to access MATRIX again

Use the arrow keys to highlight   EDIT

Use the arrow keys to highlight [B]

Press ENTER

Press 3 ENTER

Press 1 ENTER

A **three by one** matrix of should now be shown on the screen

Enter the required numbers : 12 ENTER, 40 ENTER, 60 ENTER
to display the matrix

Press 2nd MODE to access QUIT

Press 2nd x^{-1} to access MATRIX again

Press ENTER

[A] should appear on the screen

Press x^{-1}

Press 2nd x^{-1} to access MATRIX again
Use the arrow keys to highlight [B]

Press ENTER ENTER to get the required result $\begin{pmatrix} 2 \\ 8 \\ -2 \end{pmatrix}$

www.ingramcontent.com/pod-product-compliance
Lightning Source LLC
Chambersburg PA
CBHW051336170526
45166CB00002B/840